Maiara Castro

Cultural Tourism in Porto Seguro-Bahia

Maiara Castro

Cultural Tourism in Porto Seguro-Bahia

Discursive Aspects

ScienciaScripts

Imprint
Any brand names and product names mentioned in this book are subject to trademark, brand or patent protection and are trademarks or registered trademarks of their respective holders. The use of brand names, product names, common names, trade names, product descriptions etc. even without a particular marking in this work is in no way to be construed to mean that such names may be regarded as unrestricted in respect of trademark and brand protection legislation and could thus be used by anyone.

Cover image: www.ingimage.com

This book is a translation from the original published under ISBN 978-613-9-67675-0.

Publisher:
Sciencia Scripts
is a trademark of
Dodo Books Indian Ocean Ltd. and OmniScriptum S.R.L publishing group

120 High Road, East Finchley, London, N2 9ED, United Kingdom
Str. Armeneasca 28/1, office 1, Chisinau MD-2012, Republic of Moldova, Europe
Printed at: see last page
ISBN: 978-620-7-70494-1

PRESENTATION

This book is made up of 5 sections:

The first section consists of the introductory part of the book.

The second section deals with tourism and culture, conceptualizing them and discussing the main characteristics intrinsic to these subjects, forming a relationship between the two.

The third section deals with tourism marketing and destination promotion, listing the strategies used by marketing to publicize and promote a tourist destination.

The fourth section presents the analysis of the Porto Seguro-Bahia advertisements from the perspective of French Discourse Analysis (DA), in order to verify the symbolisms imprinted on the advertisements promoting the municipality.

Finally, the last section aims to present the final considerations on the subjects covered.

SUMMARY

I dedicate this work to my mother

Maria Ilza Conceicao Castro and my grandmother **Amelia Gregorio**

In memoriam.

And to my father **Geraldo de Matos Castro and** brothers **Vinicius Conceicao Castro** and **Maic Conceicao Castro.**

Because they are my reason for living and existing.

ACKNOWLEDGMENTS

To my family, friends and everyone who helped me pursue my studies: Maria Ilza (mother) and Amelia Gregorio (grandmother) in memorian, Geraldo Castro, Vinicius Castro, Maic Castro, Cristiane Soares, Luiza Soares, Rosalia Castro, Iracy Castro. My sincere thanks go to my husband Fernando da Cruz Lima and my friends Nilda Carvalho and Maria Do Carmo Aguiar.

To my advisor, Professor Moises Viana, for whom I have great admiration for his understanding, solidarity and wisdom.

The University of the State of Bahia, the degree course in Tourism for providing the wonderful experience of understanding tourism.

I would like to thank FAPESB for the Scientific Initiation Scholarship which helped in my academic development. Finally, my thanks to those who imagined this utopia with me.

1 INTRODUCED

In this study, Discourse Analysis (DA) of advertisements for cultural tourism will be carried out on advertisements for the Costa do Descobrimento, which is located in the far south of Bahia and includes the municipalities of Porto Seguro, Belmonte, Santa Cruz Cabralia, Eunapolis and Itabela. However, the focus of this study will be on the city of Porto Seguro-BA, located 722 km from Salvador and 63 km from Eunapolis on the BR-367 highway.

The Ministry of Tourism has classified 65 tourism destinations in Brazil. Porto Seguro is one of the five inducing destinations in Bahia, along with Salvador, Lengois, Mata de Sao Joao and Marau (SETUR, 2013). This destination's main economic activity is tourism and it has numerous natural and cultural attractions. Historically, the municipality stands out for being the birthplace of Brazil, the place where the country's history began.

In its entirety, the research seeks to analyze the discursive corpus of advertisements aimed at cultural tourism through Discourse Analysis (DA) in the city of Porto Seguro, understanding the importance of marketing the place as a way of publicizing both its natural and cultural attractions. In this way, the study analyzes the discourses used in the advertisements and the resources used to sell the tourist destination.

If we look at the most diverse civilizations, we will find a variety of cultural expressions, since each group develops its own way of seeing the world. Brazil has a rich variety of cultural expressions that have great sustainable economic potential (DIAS, 2006). In this way, Geertz (1989) understands culture as a set of meanings embedded in the manifestations, customs and values that man creates in order to survive. However, it is necessary to interpret these relationships that materialize in discourse. Culture becomes evident in manifestations, whether cultural or contextual.

As one of the most practiced activities in the world, tourism is considered to be one of the activities that generates the most profits for the world economy, involving crucial players in its development such as: the public sector, the private sector, NGOs and the community, encompassing social, economic, political, environmental and cultural aspects (DIAS, 2006). In Brazil, tourism is in its second phase in which, since the opening up of the economy and the stabilization of democracy, it has found ideal conditions for growth (DIAS, 2006). In this way, Wahab (1991 apud DIAS, 2003, p. 30) defined tourism as: "an intentional human activity that serves as a means of communication and as a link for interaction between peoples, both within the same country and outside the geographical limits of countries". Cultural tourism has been developing more and more, since a large part of the motivations for practicing tourism are cultural.

Porto Seguro is considered to be the birthplace of Brazil, as the first expeditions commanded by Pedro Alvares Cabral arrived there in the 1500s. Thus, it is from historical vestiges such as houses, churches, the Landmark of Possession and others that the history of Brazil is told and the city promoted, attracting more and more visitors to learn about the country's history. Cultural tourism is understood as tourist activities related to experiencing the set of significant elements of the historical and cultural heritage of cultural events, valuing and promoting the material and immaterial goods of culture" (BRASIL, MTUR, 2006, p. 13). It is therefore important to emphasize that the Costa do Descobrimento is the cradle of Brazilian history and culture.

One of the strategies used to publicize a product's potential is advertising, which is an essential tool in propagating and selling the product. Tourism also uses this tool to publicize destinations and consequently increase tourist demand. Advertising is communication with certain characteristics, aimed at promoting a product, monodirectional, intentionally formatted in a Marketing plan (PEREZ, 2006). Thus, for Kotler (2000 apud LOHMANN, 2008, p.151), marketing is "a function and a set of processes that involve creating, communicating and delivering value to customers, as well as managing relationships with them in a way that benefits the organization and its stakeholders".

These tools are constantly used by the municipality of Porto Seguro-BA to publicize its attractions and attract more and more tourist demand, imbuing the advertising discourse with a multitude of meanings. In this way, we hope to understand how the advertising discourse on cultural tourism in Porto Seguro-BA works.

The main objective of this work is to evaluate the discursive aspects of cultural tourism advertisements about the city of Porto Seguro, Bahia. The aim is to report on the characteristics of the discourses used to promote the destination, to analyze the characteristics of the discourse of the advertisements through DA and to create a discursive corpus in relation to the advertisements used.

The analysis of Porto Seguro's advertising is based on the theoretical model of Discourse Analysis (DA), critically evaluating the structures of the discourses used in the advertisements that seek to highlight the positive characteristics of this destination, through symbols and ideologies produced in parallel. Rhetoric makes it possible to see theoretically which elements generate persuasion. The rhetorical world is that of opinions, acting on human issues of a universal nature, and applies to any area of knowledge regardless of the technical particularities of science (PERES, 2006). Discourse Analysis does not seek the "true" meaning, but the real meaning in its linguistic and historical materiality (Orlandi, 2010).

Methodologically, this research was based on a case study of the municipality of Porto Seguro on the Discovery Coast. In order to analyze the corpus of discourse in the advertisements, we will use the theoretical basis of Discourse Analysis (DA), which will evaluate the meaning and characteristics of the discourse used in the Porto Seguro-Ba advertisements. To collect the data, *on-site* visits were made between May 17 and June 27, 2012 in the region where the advertisements could be located.

The data was analyzed using the standards and techniques of Discourse Analysis (DA), through paraphrase and polysemy. In other words, the analysis of linguistic structures and the sense of interlocution of discourse and the internal plurality it contains (ORLANDI, 1996).

This study was divided into three chapters, namely Chapter I, in which the conceptualization and characterization of the phenomenon of tourism was carried out through the use of literature that refers to this activity, contextualizing it with aspects of culture and addressing elements that are associated with them.

Chapter II discusses Marketing, its actions and destination promotion, describing its main elements, such as the Marketing Mix and communication, and highlights the actions of Marketing to promote a tourist destination.

Later, in Chapter III, we explained the French Discourse Analysis Methodology and analyzed the advertisements for Cultural Tourism in the municipality of Porto Seguro-BA.

CHAPTER 2

TOURISM AND CULTURE

Travel has been a crucial element in society since the earliest days of civilization. With the evolution of society and the consolidation of globalization, this phenomenon has undergone transformations, especially in terms of the motivations for travel. What was initially used as an indispensable factor in satisfying physiological needs has now been turned into entertainment and leisure (TALAVERA apud GANDARA, 2011).

> The movement of people between places is an ancient phenomenon in human society, becoming a significant element after technological advances in the 19th century. This phenomenon was given the name of tourism, leading thousands of people to travel for the most varied of reasons, which in contemporary times is perceived for its socio-cultural and economic impacts on the various peoples at a global level (TALAVERA apud GANDARA et al. 2011, p.3).

Thus, the movement of people gave rise to one of society's most influential phenomena, tourism. Thus, according to Rose (1999 apud Alberto, 2009, p.4), "The word *'tourism' (tourism, touris* - comes from *tour,* which means circuitous journey; outward and return travel); has French origins and was first used in 1760 in England".

With the spread of globalization around the world, tourism, like culture, has undergone many transformations, in a constant process of unfolding in which old elements have been lost and new ones consolidated.

> Globalization is one of the elements marking the revolution in modern society, and culture is also suffering from these transformations. And so culture, a "rediscovered" element (in many cases with "inventions of traditions") becomes an element to be exploited by tourism (ALBERTO, 2009, p. 4).

In this sense, it is understood that with the new technological discoveries and the transformations based on globalization, a historical-cultural process, culture began to suffer from the transformations imposed by the intensity of the multiplying and homogenizing effects of this phenomenon, becoming an element exploited by tourism.

According to MTUR (2008 apud GANDARA et al. 2011, p. 3), "tourism has an intimate relationship with culture, with the formation and transformation of identity, as well as the maintenance and preservation of the historical, natural and cultural heritage of certain locations". In this way, it is understood that tourism and culture are elements disseminated in society that are capable of promoting the modification of environments and the propagation of society's heritage assets.

Tourism is an activity capable of transforming the economic reality of many places and, in the process, culture also undergoes changes. In this way, Avila (2009, p. 19) states that: "Tourism, inserted in the capitalist context, in the consumer society and in the cultural industry, can commodify cultures, making them objects for consumption by visitors, thus causing damage that can be irreversible." In this way, it is understood that tourism and culture have proven to be important elements of society, promoting the development of many places, in which the relationship between these two social strands points to the transformation of culture into a marketable product, in which the cultural assets of a given community become an object of consumption and appropriation.

According to Avila (2009), the interest in getting to know the culture of others and experiencing cultural experiences is outlined in cultural tourism. In this sense, it is understood that cultural tourism is the appropriation of tourism activity over culture.

With the changes brought about by globalization, such as the advance of technology, tourism has been undergoing various transformations and reinventing itself. The search for the new has become a necessity, Avila (2009, p.27) mentions that "the search for authenticity and differentiating elements has become a market necessity, since authentic culture can be used to create tourism products". In this sense, the differentiation of cultures becomes essential for the diversification of the market, characterizing it as multifaceted and capable of pleasing the most different desires and needs.

From this perspective, which governs cultural diversification, tourism appropriates the most varied characteristics of culture, which means that the cultural segment within tourism activity has sub-segments resulting from segmental divisions.

The relationship between culture and tourism is a reality, and it is becoming increasingly difficult to dissociate them. "The tensions and contradictions arising from the relationship between visitors and hosts increase the need to plan this encounter" (AVILA, 2009, p.27). In this way, the union between tourism and culture involves actors participating in the activity who, as a result of the development of the tourism process, may suffer the effects caused by tourism.

The relationship between these elements is a paradoxical process, with a multiplicity of effects, both positive and negative (AVILA, 2009). In this way, tourism planning should seek, through the development of strategies, to focus on the peculiarities that permeate the culture of a given community, preserving it.

2.1 TOURISM AND ITS RELATIONSHIP WITH CULTURAL ASPECTS

Tourism is one of the most influential global phenomena, developing regions in different parts of the world. Its characteristics include promoting the dissemination of data about certain regions, countries, cities and communities, diagnosing potential for economic development through the activity, triggering new expectations in the socio-cultural and economic spheres, as well as promoting socialization (DIAS, 2006). In this way, it can be understood that tourism has been gaining worldwide prominence over the years as an activity that promotes social integration and economically develops many destinations. In this sense, tourism has a multitude of aspects and, with this, a multitude of definitions by bodies related to the activity and by authors who seek to understand the totality of the activity. Thus, Dias (2003) defines tourism as activities carried out by people traveling or staying in places other than their usual ones, for a specific period of less than a year, for leisure, business or other reasons. Furthermore, tourism involves motivation, seasonality, goods, services, tourist agents, structure and community involvement.

The World Tourism Organization (1994, p.38) defines tourism as "the activities that people carry out during their travels and stays in places other than their usual environment, for a consecutive period of less than a year, for leisure, business or other purposes".

Globalization is a societal phenomenon taking place on a worldwide scale. This process consists of an integration of economic, social, cultural and political aspects between different countries. Globalization came about as a result of innovations in technology and the means of communication. This has changed the relationship between processes, shortening distances between places, facilitating the industrial process and other sectors of the economy (DIAS, 2006).

The historical economic process shows that countries' economies are linked to each other, and when an event affects one country's economy, it immediately affects another's and vice versa. In recent years, the world economy has undergone obvious changes and the economic focus that used to be concentrated on industry has shifted to the services sector.

> The effects of industrialization on society were so significant that the period came to be known as the industrial era. With its decline, the new phase, although it is not well characterized, as it is still ongoing, has been called, by various authors, post-industrial or post-Fordist, in which the axis of economic activity shifts to the services sector (DIAS, 2006, p. 2).

In this sense, it is known that services have peculiar characteristics and Kotler (2000, p. 448) defines them as: "any essentially intangible act or performance that one party can offer to another and that does not result in ownership of something. The execution of a service may or may not be linked to a physical product". In this sector we can include tourism, which is currently considered an important product of the world economy, gaining increasing prominence as one of the most favorable alternatives for economic development in many countries, considering that the activity is one of the fastest growing worldwide.

> Tourism is the fastest growing sector of the economy, surpassing traditional sectors such as the automobile, electronics and oil industries, which are also considered the world's main economic activity (DIAS, 2006, p. 3).

Tourism is a typically transformative activity which, when applied to a place, promotes countless changes in the production process, adapting it to receive tourists.

> Tourism has a particular relationship with the territory in which it is practiced, because the places chosen for tourism are modified in a process of touristification, which occurs to meet the demand of tourists, who need, during their stay, specific infrastructure, services and equipment that make their stay pleasant (DIAS, 2006, p. 15).

The tourist activity is a set of interdependent elements that form an organized whole with correlated characteristics making up a system, this system is made up of subsystems that are inseparable from the practice of the activity.

> Tourism, in the language of General Systems Theory, should be considered an open system which, as defined in the structure of systems, allows the identification of its basic characteristics, which become the elements of the system (BENI, 2003, p. 44).

This phenomenon is a wide-ranging activity that manages to connect a multitude of factors that revolve around man and uses these resources according to the development of the activity, generating modifications and adaptations. This activity manages to use both natural assets such as waterfalls, rivers, beaches, lakes and so on, as well as heritage assets such as museums, historic houses and the cultural manifestations of a people, making them marketable products.

> Tourism is an activity that manages to involve all aspects of human existence and its natural surroundings, as well as transforming both natural resources and tangible and intangible cultural heritage into a marketable product (DIAS, 2006, p. 14).

The tourist product has characteristics that are very common to other activities in the service sector, while other characteristics are unique and specific to them. The tourist product is made up of goods and services and is also the result of the actions of the agents involved in the activity.

> Tourism arises from a set of activities of a heterogeneous nature that prevent it from becoming an autonomous science and independent specific techniques. It does not have a disciplined and rigid organization, nor does it have its own methodology (ANDRADE apud DIAS, 2006, p. 11).

As tourism is one of the activities that most influences the world economy, it has aroused and continues to arouse the interest of both federal, state and municipal governments and companies, as it offers significant income and countless business opportunities.

> In tourism, one can imagine, a priori, that both the state and business sectors have profit as their real objective. The state expects a surplus on the balance of payments in the specific account, due to the inflow of foreign currency, and the companies that operate in the sector also scale the provision of their services according to the profitability of the necessary investments (BENI, 2003, p. 25).

Based on this interest, governments have intervened to promote tourism and maximize the economic and social benefits of this activity through strategic planning and initiatives. However, despite being one of the most important and profitable activities in the current global economy, favoring countless communities economically and socially, this activity is still placed in the background in the planning and preparation of public policies by the government at all levels.

> Although tourism is one of the most significant expressions of today's globalized economy and one of the dynamic players in the process of world integration, it has been relegated to a secondary level in the drawing up of public policies at all levels of state organization (federal, state and municipal) and disregarded as a tool for development (DIAS, 2006, p. 6).

It is therefore understood that the Brazilian state does not believe in the development potential of tourism, taking into account other economic factors. This disbelief in the activity substantially reduces the possibilities and abilities of tourism to reduce inequalities, whether social or economic.

This phenomenon as a system enables the interconnection of the elements that are part of tourism, resulting in social and economic development achieved through the multiplier growth in the flow of internal and external demand and the generation of investments by private companies and the government (BENI, 2003).

According to Dias (2006), tourism is an activity that, when implemented, has both positive and negative effects, whether in the social, economic, cultural, environmental or political spheres. From an economic perspective, tourism, if well planned and monitored, becomes one of the possibilities for economic and social development and can reduce social inequalities and poverty, generating employment and income. For this activity to be successful, it is necessary to implement public policies and planning with the participation of the main players involved: government, NGOs, local communities and the private sector.

> The way tourism is implemented in a given territory is directly related to the policy adopted by the local political administration. It can be implemented quickly and in a disorderly manner, which generates benefits in the short term, but eventual damage and unsustainable development in the medium and long term (DIAS, 2006, p. 16).

In this sense, it is understood that strategically prepared planning is an indispensable element that can prevent the generation of negative effects of tourism that compromise the sustainable development of a community.

The potential of tourism as a complex phenomenon that unfolds in economic, social, political and cultural dimensions develops an interesting reflection on places, i.e. tourist destinations, countries, regions and cities (VALLS, 2004). In this way, tourism is understood as a wide-ranging phenomenon that encompasses the social, economic, political, environmental and cultural aspects of society, promoting positive and negative changes in these areas.

The activity can be considered one of the factors of development and generator of the intensification of socio-cultural integration processes. As it presents a paradoxical context, tourism is full of uncertainties in which plurality and contradiction are striking characteristics (SPINOLA, 2008). Spinola (2008, p.1) states that:

> [...] apparently there are no absolute certainties. On the contrary, the duality and contradiction between the various trends manifested, as well as the complementarity and cause-consequence relationship between the same trends, and between them and the actions implemented by the public and private spheres, form an inseparable whole that is difficult to analyze.

Based on the conception and appreciation of the importance of tourism, local communities try to develop the activity in a way that seeks to maximize the benefits of tourism in the locality. This tourist mobility of autochthonous communities can be analyzed as a tourist culture.

> Tourism is fundamentally cultural in nature, as it is a process of continuous interaction between different communities that occupy different socially constructed spaces and which, because of this diversity, become attractive to the other - the

> tourist, who travels to see new places, to rest, to relax in an environment different from the one where they live (DIAS, 2006, p. 1).

In this sense, tourism has been crucially established on the basis of culture, since the cultural diversity of society provokes in man the curiosity to know the new, to visit new places, to establish contacts with different cultures, fulfilling the need to interact. Nowadays, tourism has become a determining factor in the process of economic and social development, bringing many benefits to the most diverse regions and consequently providing local development. Cultural diversity is one of the main factors motivating tourists to visit different communities in order to exchange knowledge, acquire new experiences and get to know something new. The unfolding of tourism activity brings with it the cultural exposure of the cultural homogeneity of a particular community.

> Tourism is inextricably linked to culture, and this is becoming more evident at the beginning of this century, due to the growing awareness that cultural diversity is the main ingredient for development, which has proved to be extraordinary, to the point that in many regions tourism has become the main economic activity, responsible for generating employment and income (DIAS, 2006, p, 1).

In view of the relationships caused by tourism, it is important to address the interactions between the autochthonous community of a place and tourists, who establish intercultural contact that intensely causes changes in the cultural exchange of the agents who interact during the process of tourist activity. This intercultural contact will always bring about changes in societies because:

> societies that allow their members extensive contact with other societies can expect to change more quickly and become more complex than societies whose members have little contact outside their local groupings. The greater the scope for novelty to which people are exposed, the more likely they are to adopt new forms. Contact between societies is the greatest determinant of culture change (DIAS, 2006, p. 23).

According to Dias (2006), during a visit tourists tend to behave differently from the way they behave in their place of origin, since they are out of their natural daily lives, living in a different place and with different people they are not used to living with. Another important factor *is* the culture shock caused when tourists visit cultures different from their own.

> Tourists generally show a huge variety of reactions to their travel experiences. Many visitors, as soon as they become aware that they are far from home, feel liberated from their normal inhibitions and start to act like different people, adopting social behaviors that are far removed from those adopted in their daily lives (DIAS, 2006, p. 22).

The level of contact with the other culture can also be a determining factor in the degree of cultural knowledge obtained and the interrelationship between the actors involved, which will be established based on the tourist's motivation for visiting the destination, since some motivations lead to greater contact with the culture than others.

One of the main characteristics of tourism is that it is changeable, causing both beneficial and harmful changes in the receiving community. When it is well planned and structured, it can provide services and infrastructures, generate jobs and income, thereby strengthening the local economy, preserve cultural and natural heritage, among other things, and have negative effects when it is not strategically planned and maintained, such as: making the site susceptible to the development of sex tourism, exploitation of labor, consumption of licit and illicit drugs, loss of identity and authenticity of the local community's culture, distortion of values, among others (DIAS, 2006).

As Trigo (2004 apud AVILA, 2009, p. 69) points out,

> Tourism is a phenomenon that not only creates jobs, taxes and development. If poorly planned and implemented, it is a factor in pollution, social exclusion, income concentration, an increase in prostitution, an increase in child sexual exploitation and compromises in poorly designed projects.

In this way, tourism can be interpreted as an ambiguous aspect and its effects will depend on the management of the planning and implementation of the activity, taking into account the factors affecting the community. As such, tourism stands out because it is an activity that is transformed by changes in society, and it can have both positive and negative effects.

> Tourism is a socio-cultural phenomenon with profound symbolic value for the people who practice it. The tourist subject consumes tourism through a tribal process of communion, fulfillment and witnessing, in a space and time that is both real and virtual, as long as it is possible to live together and be present. The symbolic value permeating the tactile communication of this phenomenon is reproduced ideologically when tourists share the feelings reproduced by fun, and when there is the possibility of materializing the imaginary, sometimes individual into societal (BENI, 2003, p. 41).

The process of building tourism arose from man's need to travel and develop contacts with other peoples and cultures, thus constituting an important social phenomenon.

2.2 CULTURE AND ITS IMPORTANCE

Culture means a set of meanings that man creates in order to survive. In order to do so, it is necessary to interpret these relations that materialize in discourse. The question of culture then becomes evident in the specific and contextual cultural manifestations (GEERTZ, 1989). In this way, we can see that customs, clothing, languages and religions are all part of a historical and constructive process in society, so culture is everything that has been created by humanity throughout its existence. All this legacy awakens a sense of belonging to a particular historical social group that develops actions related to its culture. Human beings do not live according to their instinctive needs, but with the ability to think about reality, constructing an infinite number of meanings for the processes that surround them.

The cultural context is reflected in the social practices that are common to a society: the simple act of eating with cutlery, combing one's hair, dancing, festivities, clothing and painting one's nails are all manifestations that are part of a cultural conjuncture (DIAS, 2006).

> For each group, there is a social form, a religious agenda, a construction of symbolic languages and many other mediations that have allowed the space to be gradually transformed into an accumulation of very characteristic forms which, today, clash with the post-modern world, in a relationship that is not necessarily harmonious, but with possibilities for appropriation and profitable management through various activities, such as tourism, for example. Thus, it is considered that architectural forms, clothing, language, folklore, typical foods, religious manifestations are representations that should be thought of as the legacy left by peoples who preceded the current social groups and which, even with new trends and innovations, are maintained, perpetuating the legacy passed down from generation to generation (PORTUGUEZ, 2004, p. 24).

Hall (2005 apud VIANA, 2011, p. 40) corroborates this,

> highlights the universality of culture in people's lives and in society, from the most everyday activities to the most complex human activities, but which is in conflict, given that it is a relationship of contrasts that is written in the transformations posed by modernity and capitalism.

In this sense, cultures are elements of each society, each with its own values, identity, historical and cultural aspects. This changes according to time and space, the perception of each community and the aspects that govern it.

Thus, the United Nations Educational, Scientific and Cultural Organization (UNESCO) sought to define cultural goods and services, since these are elements inserted into culture. At the Convention on Cultural Diversity, which took place in Paris, UNESCO (2005 apud DIAS, 2006, p.22) defined cultural goods and services as,

> All those goods, services and activities that embody or give rise to cultural expressions and that have the following characteristics: they are the result of human work - industrial, artistic or craft - and their production requires the exercise of human creativity; they express or convey a certain symbolic meaning, which endows them with a cultural value or significance that is distinct from any commercial value they may have; they generate, or may generate, intellectual property, regardless of whether or not they are currently protected by existing intellectual property

legislation.

In this way, cultural goods and services are elements inserted in the social context, which are generated from the processes that move society. Dias (2006) believes that culture is an important element in the conceptualization of cultural heritage.

> Cultural heritage is currently considered to be a set of tangible and intangible assets that have been bequeathed by our ancestors and which, from a perspective of sustainability, will have to be passed on to our descendants, with new content and new meanings, which will probably have to undergo new interpretations in accordance with new socio-cultural realities. Cultural heritage is made up of tangible and intangible elements - traditions, literature, language, handicrafts, dance, gastronomy, clothing, religious manifestations, historical objects and materials, architecture, etc. - both past and present, which together characterize a social grouping, a people, a culture (DIAS, 2006, p. 67).

In 1972, the UNESCO World Heritage Convention (2013 apud BARRETO, 2000, p. 12) defined cultural heritage as,

> Monuments: monumental works of architecture, sculpture and painting, elements or structures of an archaeological nature, inscriptions, caves and combinations of these that have a value of universal relevance from the point of view of history, art or science; group of buildings: separate or connected groups of buildings which, because of their architecture, homogeneity or location in the landscape, are of universal significance from the point of view of the history of art or the sciences; sites: works made by man or by nature and man together, and areas including archaeological sites which are of universal significance from the point of view of history, aesthetics, ethnology or anthropology.

The National Historical and Artistic Heritage Institute (IPHAN, 2013) defines Cultural Heritage reporting that:

> It is not restricted to isolated official buildings, churches or palaces, but in its contemporary conception extends to private buildings, urban areas and even natural environments of landscape importance, including images, furniture, utensils and other movable goods (p. 1).

Thus, it is understood that cultural heritage *is* made up of the goods left behind by ancient societies, historical aspects, historical and cultural manifestations, organized into tangible and visible tangible heritage and untouchable and felt intangible heritage.

Dias (2006) points to the emergence of social arrangements based on the variability of interactions between cultures caused by the phenomenon of globalization. From this perspective, the author points out that

> Developing culture in contemporary, multicultural and densely interconnected societies cannot consist of privileging one tradition or simply preserving a set of

> traditions unified by a state as a "national culture". The most productive development is that which values the richness of differences, fosters communication and exchange-internally and with the world-and contributes to correcting inequalities (p. 21).

Valuing and respecting cultural diversity becomes a determining factor in understanding the differences imprinted on each community.

2.3 CULTURAL TOURISM: A SEGMENT BASED ON THE MEETING OF CULTURES

Tourism segmentation has emerged as a planning strategy for the activity, providing for the most varied motivations and demands, boosting the flow of tourists to destinations.

Tourism modalities are changeable and undergo subdivision processes, consequently increasing the number of segmentations or regroupings that join segments together to form new typologies, as is the case with the business tourism and events segments, which were originally distinct modalities. Lohmann (2008, p.172) states that "the segmentation of the tourism market increases daily with segments being subdivided, while others regroup and become a new segment".

> Segmentation is a process that should always be taken into account in tourism, as it facilitates the marketing of the tourist product, allows for a more professional formatting of the product with the hiring of specialists who will enable the continuous improvement of the attractions favored by the locality, which would not be possible if one decided to work with a very wide range of cultural tourism segments. Identifying a priority in terms of cultural tourism makes it possible to attract sectors that complement the main attraction, so as to make it possible to combine interests (DIAS, 2006, p. 56).

From this, it can be understood that segmentation is an important process that facilitates the marketing prerogative, making tourism easier to apply.

Cultural tourism encompasses a variety of cultural aspects that can include folklore, cultural and religious events, museums, historic houses, festivities, traditions, dangas, architectural monuments, historic sites, cultural events and other representations that reflect the aspects that identify the historical and cultural processes of a given place.

According to ICOMOS (1976 apud Dias, 2006, p.39), in the Charter of Cultural Tourism, cultural tourism ë:

> That form of tourism which has as its object, among other things, the knowledge of monuments and historical-artistic sites. It has a really positive effect on them as much as it contributes to their maintenance and protection in order to meet its own objectives. This form of tourism in fact justifies the efforts that such maintenance and

> protection require of the human community, due to the socio-cultural and economic benefits it brings to the entire population involved.

In order for cultural tourism to be defined in its entirety, it is essential to encompass and analyze the elements that make up the segment, such as: tourists' motivations, experiences and experiments, interrelationships, among others.

Cultural tourism can help preserve cultural heritage by revitalizing identity, life experiences or even enhancing unique cultural aspects (MTUR, 2008). It is therefore important to emphasize that cultural tourism can be explained as the tourist segment that enables tourists to come into contact with the cultural heritage and socio-cultural process of the autochthonous community. This type of tourism can function both as an instrument for restructuring the identity and authenticity of the culture of the resident population and as a tool for local economic development. In this sense, for Dias (2006, p. 46) "cultural heritage is the essence of cultural tourism, the great motivation for tourists to travel and valuable cultural capital for communities, as it represents a tourist product which, if well managed, can last indefinitely".

Cultural tourism has become one of today's most significant segments, attracting tourists whose motivation is to get to know the culture of others, to learn about the past, to rediscover and affirm their cultural identity in the face of globalization (DIAS, 2006). These tourists generally have peculiar characteristics.

> It should always be borne in mind that cultural tourists, as a niche market, form a particular type of person, generally with a high level of environmental awareness, a broad political vision and an interest in different cultures (DIAS, 2006, p. 36).

Thus, it can be concluded that interest in the cultural aspects of a place is shown by people who are highly educated, seek past and scientific knowledge, are willing to maintain relationships with locals, value quality service and prefer to stay in exquisite places, appreciate authenticity, are willing to spend money, among other characteristics. These are considered direct customers of tourism. On the other hand, Dias (2006) points out that tourists who visit a destination for a variety of reasons other than cultural ones are called indirect customers, since they don't visit in order to obtain scientific knowledge, but merely to have a superficial understanding of what is presented to them during their stay at the destination.

According to Talavera (2003 apud VIANA, 2011, p. 56),

> There are many motivations for the cultural tourist, for example, ranging from the

cultural experience as a goal or as a mere detail of a relaxing trip: middle class people who don't spend a lot, who are eager for knowledge, concerned about nature and the preservation of cultural heritage, an identity that they revitalize with their life experience.

In this way, it is understood that tourists who travel motivated by culture seek to absorb knowledge about the culture of others, to experience customs, to establish contact with the community in order to understand the peculiarities that make up the culture of a given location or even to live for leisure.

As it is a diversified activity, there can be many different motivations for tourists visiting a given destination, even so, those who have a different motivation from the one seeking cultural knowledge indirectly maintain contact with the culture they are visiting. Dias (2006, p. 37) explains that: "These tourists who don't seek out cultural heritage as their first option are what we can call indirect clients of cultural tourism." This means that, although many of the motivations are different from those for cultural tourism, this type of tourism can be combined with other tourist segments that make it possible to carry out a range of other practices, such as ёйкс tourism, which is a segment that has affinities with cultural tourism, educational tourism that makes it possible to increase the visitor's awareness of local socio-cultural aspects, among other activities.

> [...] Access to historical knowledge through the formulation of the concept of heritage involves various elements combined with political situations, economic values and personal attitudes. Heritage thus becomes value, capital, and must be managed as such (COSTA, 2004, p. 1).

Visiting places with architectural monuments and/or historical sites allows visitors to awaken feelings that connect the moments they have lived with their past, allowing them to recover their authentic identity.

> Going to a historic site represents a journey back in time that allows the viewer to have an experience of emotional and physical contact with countless pieces of equipment that date back to their past, giving more meaning to the history of their life, their family, their community and even their country (PORTUGUEZ, 2004, p. 4).

On the other hand, visiting historical sites can also be motivated by the desire to learn about the history of other places, allowing tourists to try out new experiences and report on their passage through the most diverse remnants of the culture of other peoples. These experiences, whether they be contact with the past or learning about the history of new cultures, can only be experienced at the time of the visit (PORTUGUEZ, 2004).

> These are just a few of the factors that give historical and cultural tourism its strong appeal today, as well as a very important operational factor: historical sites, with their architectural forms, their culture, their aesthetic manifestations and many other categories of materialization of/in place, can only be experienced *in loco, that* is, no matter how much they reproduce places with the evolution of technology, information and engineering, the act of actually being in a given place is an element of great symbolic value that attributes new values and uses to very specific environments, where ruins, buildings, caves, streets and houses gain emotional meanings that strengthen their identities (PORTUGUEZ, 2004, p. 5). 5).

Globalization, through its homogenizing tendency, has increasingly affected the insertion of consumer desires, which every day increases people's interest in getting to know new places, new cultures and new people, making cultural tourism one of the most significant expressions of today. Based on this principle, it is believed that cultural tourism will suffer from the process of massification over time (DIAS, 2006).

> The trend towards the massification of cultural tourism is inevitable, as there is an increase in people's interest in getting to know their own surroundings, the history of their local community, their country and of humanity itself, as they see themselves increasingly integrated into this imagined global community that is emerging due to the intensification of the globalization process. It is believed that in the future, heritage sites will have difficulties adapting to the new massification of the segment, since tourist demand will grow respectively (DIAS, 2006, p. 42).

Cultural motivation enables the acquisition of scientific, cultural and historical knowledge in a way that allows the consumption of cultural heritage and contact with the indigenous community. In order for this heritage to endure and cultivate the memory and identity of a people, it is necessary to raise awareness of the importance of these assets for humanity, as well as their conservation. In this sense, Portuguez (2004, p. 8) states that: "Heritage refers to the people, the origins and the history of a community. The need to preserve heritage in order to reinforce the identity of people and places in the first place and, if there is tourist potential, its use, its recreational use."

It is through the activity of tourism that many architectural monuments attract the attention not only of tourist demand but also of the authorities who seek to restore, recover and preserve important heritage sites.

> Tourism, whether for cultural or mass reasons, has undergone significant development over the last decade and is becoming a key player in the life and also in the urban and architectural recovery of important historic sites, inducing processes of rehabilitation and reuse of monumental buildings, as well as improvements to public spaces and cultural infrastructures and facilities (PORTUGUEZ, 2004, p. 35).

In a process in which material heritage is transformed into marketable products, it is essential to apply measures that can preserve and consequently guarantee the continuity of the heritage legacy.

The United Nations Educational, Scientific and Cultural Organization (UNESCO, 2006, p. 1) defines Intangible Cultural Heritage as:

> [...] the practices, representations, expressions, knowledge and techniques - with the instruments, objects, artefacts and cultural places associated with them - that communities, groups and, in some cases, individuals, recognize as an integral part of their cultural heritage.

The preservation and conservation of intangible cultural heritage linked to sustainable management with new social and economic expectations can guarantee the perpetuation of a society's cultural identity, memory and traditions. Limiting and defining tourist demand could be one of the measures taken to preserve heritage.

> Sustainable management of cultural heritage, a strategic resource for historic cities, requires reconciling conservation with new economic and social expectations that offer emerging functions or those that best fit their urban structure (PORTUGUEZ, 2004, p. 4).

Thus, it is known that cultural heritage is of great importance to humanity since it includes the memory, identity and culture of peoples. It is known that preserving cultural heritage is of the utmost importance, as it will allow humanity's cultural legacy to be perpetuated from generation to generation.

Thus, Viana (2011, p. 50) points out that: "Preserving intangible cultural heritage and its human aspects means resisting fragmentation, it means being fully subject to your past, present and future, it means maintaining your identity". It is thus understood that valuing and preserving become crucial characteristics for the perpetuation of a people's identity.

Cultural tourism is a form of tourism that can provide many positive experiences for tourists, offering both tangible and intangible cultural heritage as attractions, enabling them to understand and develop awareness of the preservation of heritage assets.

CHAPTER 3

TOURISM MARKETING AND DESTINATION PROMOTION

Marketing is one of the most important tools for promoting services and products. This science seeks to analyze the desires and needs of consumers in order to satisfy their needs efficiently. Marketing is about the relationship between supply (supplier) and demand (consumer). According to Palmieri et al. (apud BOTELHO; COUTINHO, 2007 p.3), "Marketing is a social science that aims to analyze consumer behavior, seeking to satisfy their desires and needs, in order to offer products and thus make a profit".

Ja Kotler (2006, p. 4) contributes by defining Marketing as:

> an organizational function and a set of processes that involve creating, communicating and delivering value to customers, as well as managing the relationship with them in a way that benefits the organization and its stakeholders.

This marketing tool is also described by Kotler (2000, p. 25) as being: "the task of creating, promoting and supplying goods and services to customers, whether they are physical or legal persons". In this way, it can be seen that Marketing manages to take in various branches, developing a multitude of functions.

Marketing is a tool that seeks to promote customer satisfaction in order to create and promote products that meet their needs and desires. Thus, Botelho and Coutinho (2007) state that Marketing is based on two central ideas: the target market, which aims to select the market focus for the application of Marketing techniques, and the satisfaction of consumer needs, which points to the analysis of consumer needs so that they can become interested in the product or service on offer.

Attracting consumers' attention, maintaining and intensifying customer relationships, persuading consumers to buy a product, analyzing consumers' needs and desires in order to satisfy them are some of the purposes of marketing, which has now become an indispensable tool in a competitive market that increasingly craves differentiated products. One of Marketing's activities is customer perception, market analysis and segmentation, promoting advertising and sales activities, and facilitating the relationship between the supplier of products or services and the customer through communication channels. According to Botelho and Coutinho (2007), this instrument has the function of promoting analysis of the consumerist desires of a given community, proposing to satisfy their needs, whether personal or social, in order to offer them compatible products and services.

For marketing to be developed in such a way as to meet the needs of the market, it is crucial to establish efficient and effective strategic planning in order to guide the decisions to be made.

In this context, Mota (2001, p. 122) states that "Marketing planning is the basis for all marketing strategy, as it defines product lines, prices, the appropriate selection of distribution channels and decisions related to promotional campaigns".

In this sense, the author reports that the Marketing plan:

> revolves around a marketing strategy that is efficient in its use of resources, flexible and adaptable. A marketing strategy can be understood as a company's overall program to select a specific target market and satisfy its consumers through a careful balance of the elements of the marketing mix - product, price, distribution and promotion - which represent subsets of the overall marketing strategy (MOTA, 2001, p. 142).

Therefore, strategic planning is the management process that will underpin marketing strategies and consequently strategic decisions in order to make balanced use of the elements that make up this tool.

For marketing planning to be well prepared, it is necessary to take into account all the elements in a marketing environment. Vasconcelos (2009, p. 48) points out that: "Marketing environment is the marketing environment in which relationships and communication occur between the communicator and the receiver (target audience)".

From this perspective, the market is defined by Kotler (2000, p. 140) as being: "the set of all actual and potential buyers of an offer to the market." As such, it is understood that the market encompasses all those who have the power to consume. As far as marketing properties are concerned, market segmentation is seen as an important marketing tactic, since it aims to identify the common preferences of a particular group, targeting them both massively and individually in order to maximize the product's effects on the market, satisfying the needs and desires of customers. Thus, the target market for Kotler (2000, p. 141), "which is also called the server market as the qualified part of the available market - is the

consumers who have the income, interest, access and qualifications for the market offer". In this way, the target market can be understood as the market segment you want to reach.

Strategic planning aims to provide a basis for marketing strategies so that strategic decisions can adapt the product to the target market. In this way, the Marketing mix or operative Marketing is used to execute the strategic plan, as Mendonca Junior (2004, p. 28) points out, "[...] the operative Marketing or Marketing mix is focused on the design and execution of the Marketing plan, in other words, it is centered on carrying out the previously planned strategy".

3.1 OPERATIONAL MARKETING OR MARKETING MIX

Within the marketing framework, strategic and tactical planning can be developed using the marketing mix, which includes the four Ps: product, pest, price and promotion. In this sense,

> In order to manage marketing, this analysis highlights the importance of certain instruments that have been created and established to serve the target market. These include the Marketing Mix, better known as the four "Ps", which are: Product, price, pest and promotion (PEREZ , 2006, p.38).

In this way, Kotler and Armstrong (2005, apud GRANDISOLI, 2013, p. 2) report that: "Marketing mix can be defined as a set of tactical and controllable marketing tools that the company uses to produce the response it wants in the target market". In this way, it is understood that Marketing is made up of a set of factors that are combined for its application in such a way as to be analyzed as the articulation of all the elements of the mix in which they enable the necessary conditions for the commercialization of the product, thus achieving the expected objectives.

Operational marketing or the marketing mix has been used by companies to achieve their objectives in the market and by tourist destinations to attract more and more visitors. In this sense, Frantz et al. (2010, p. 4) report that:

> one of the objectives of the marketing mix is to provide a clearer view of the organization of marketing in companies. Although it was proposed in the 1960s by Jerome McCarthy to characterize the complex relationships between the product, the price, the pest and the promotion, it is still fundamental to marketing strategies and actions today.

The composite of the 4 Ps should be analyzed as a set of essential elements when it comes to the marketing strategies of a destination or tourism company. Thus, Frantz et. al. (2010) point out that "the marketing mix has been applied by tourism companies in order to serve their customers, and meet their consumption needs and desires, consequently successfully achieving individual and organizational strategies, as well as tourist destinations that seek to publicize their attractions by arousing the interest of visitors.

In this sense, Kotler (2000, p. 38) states that "the 4 Ps represent the selling company's vision of the marketing tools available to influence buyers. Thus, it can be understood that the Marketing mix variables are essential instruments that are part of Marketing strategies and actions.

Based on these considerations, the product, which is one of the main variables in the composite, according to Kotler and Keller (2006 apud FRANTZ et al. 2010, p. 5), can be considered a physical object or a service. In this way, it can be understood that the product can be tangible or intangible. Kotler and Keller (apud FRANTZ et al. 2010, p. 5) also point out that it must have variety, quality, design, features, a brand name, packaging, sizes, services, guarantees and the possibility of returns.

Based on this principle, "the product is a key element in the market offer. Marketing mix planning involves formulating an offer to satisfy the needs and desires of the target customer" (KOTLER, 2000, p. 416). The product can therefore be understood as everything that can be offered on the market. "The customer will judge the offer according to three factors: characteristics and quality of the product, mix and quality of services and appropriate price".

Pricing is one of the main components of the Marketing Mix, as Frantz et al. (2010, p. 5) report, "pricing is one of the most important components to be analyzed and determined in a business. Price determination should be based on production costs, administrative costs and the cost of selling the product on the market". In this way, it is understood that the price must take into account the value given by people, in other words, the value they attribute to the product, because the price is determined by the market.

Promotion is the act of marketing the product and is defined by Kotler (2001 apud FRANTZ et. al. 2010, p. 6) as: "what leads to the sale of products and promotes the aTrayë mix of advertising, publicity, public relations and others. Promotion is the language link between the manufacturer and the consumer." In this conjecture, it is understood that promotion is the variable that makes the link between the destination or the company and the customer. Promotion can be seen as the sale and dissemination of the product through the media, Frantz et. al. (2010, p. 6) further states that: "Promoting a product consists of marketing it by creating a communication link between the company and the consumer market, through advertising on television, billboards, radio, trade magazines, newspapers and the internet."

Finally, the pest variable is defined by Kotler (2001 apud FRANTZ et. al. 2010, p. 6) as that which "consists of coverage, quantity, quality of distribution channels, logistics, points of sale, stock levels, packaging, transportation, among other items that guarantee adequate market service". Therefore, it can be seen that the plague or distribution point are the markets in which this product is marketed, which encompasses everything from distribution channels to product promotion and communication.

3.2 MARKETING AND ITS RELATIONSHIP WITH COMMUNICATION

An integral part of marketing is communication, which is associated with strategic decisions and, in order for this to be properly planned, it is crucial to understand consumer perceptions. As Al Ries and Jack Trout (apud VASCONCELOS, 2009, p. 23) point out, "Marketing is a battle of perceptions - not products". Thus, it is understood that Marketing actions seek to analyze, measure and master the concepts that are inserted in the market in congruence with consumer perceptions of a given product, acting respectively as a promoter of ideologies where strategic decisions and communication are used as essential processes to achieve the objectives set.

Communication is one of the processes within the Marketing framework that is integrated with tactical and strategic decision-making, promoting the achievement of the desired objectives. According to Ribeiro (apud VASCONCELOS, 2009, p. 19) "there is no communication planning that is not linked to Marketing - communication planning is the link between communication and Marketing". For Vasconcelos (2009, p. 21), communication "has the function of ensuring that all marketing planning, based on the needs of a particular market, reaches that market, and the company or product is communicated, becoming known to its target audience".

It is believed that, in order to achieve the expected results, it is essential to use communication as an integral part of the marketing mix in conjunction with strategically elaborated deliberations. The main objective of communication is to support the effectiveness of marketing planning so that it achieves the expected results. In this way, Vasconcelos (2009, p. 21) argues that "[...] in order for a given marketing objective to be solved by means of communication, it is necessary to evaluate all the other elements of the marketing mix".

Communication planning must take into account the applicability of the tools and how they can be used. Thus, Vasconcelos (2009, p. 19) states that: "communication planning can, for example, suggest a different way of presenting the product, an innovative use or directing the message to a new market segment that has not yet been explored". It is therefore necessary to know what Marketing is, what elements are included and what the functionality of communication is in this tool.

From this perspective, Vasconcelos (2009, p. 23) states that: "[...] Planning communication is presenting a message in such a way as to arouse in the target audience the desired perception in order to achieve the desired return". Communication planning should therefore be configured and defined after analyzing the market and, consequently, the consumer, diagnosing the problem and looking for concrete information on where the communication will act. In this way, planning seeks to establish how communication will work to reach the target audience in question and consequently obtain the pre-established return. The strategy defined to guide the measures taken in relation to the marketing of a particular company or tourist destination must include effective integrated communication planning, in which there is an integration of communication tools linked to marketing efforts.

Thus, in order for communication to enable marketing planning to reach the market, the services or the tourism product to be communicated, it must be planned effectively. Based on these considerations, Marketing can be analyzed as a fundamental tool for destinations and companies seeking to achieve success by tactically designing their actions and promoting new and established products. Tourism services or destinations find marketing to be a crucial tool for promoting their services and increasing demand.

In this context, Westwood (apud MOTA, 2001, p. 135) points out that: "the marketing organization of tourist services and/or destinations can manage its competencies and control the variables that allow a company to bring out a policy that is profitable and satisfies its clientele". Thus, it is understood that once the problem in which communication will act has been determined, one of its aims will be to analyze consumers and their reaction to certain products. Thus, according to Vasconcelos (2001, p. 41), "once communication aims to achieve a certain reaction from the selected target audience, it becomes essential to understand how consumers process their choices, evaluate products and react to certain stimuli".

Through their behavior, consumers seek to reveal their desires and needs and it is from this demonstration that marketing planning must evaluate and try to understand its target audience, enabling customer satisfaction. Thus, it is analyzed that marketing planning, as a result of analyzing perception and how the consumer absorbs market information about certain products, will make strategic decisions that can lead this consumer to buy that product.

According to Vasconcelos (2009), good planning and execution of the Marketing mix to achieve satisfactory results in the market will make it easier to sell the product. It is considered that when its actions are well planned, Marketing will promote the product without having to make any effort to sell it, thus achieving its objective. Thus, the action of Marketing must precede the sale of the product and be perpetuated throughout its continuity on the market. Kotler (apud VASCONCELOS, 2009, p. 29) goes on to say that, "when marketing is successful, people like the new product, the novelty spreads by word of mouth and little sales effort is needed".

3.3 MARKETING ACTIONS IN TOURISM

Marketing can be applied to products, goods or services. Mendonga Junior (2004, p. 32) corroborates this by stating that: "The techniques used to design a marketing plan can generally be applied to any type of product, good or service". In this respect, the applicability of the techniques used by Marketing must be adapted to each market segment in order to achieve the desired results, in other words, Marketing actions must be different when applied to material goods and when applied to services. "Services have a very specific nature that clearly differentiates them from the sale of tangible goods" (MENDONCA JUNIOR, 2004, p. 33). From this point of view, it is understood that marketing, whether focused on tourism or other marketing trends, must be analyzed and scaled to the product or service in order to adapt to market needs.

Marketing techniques to promote tourist services can be applied directly or indirectly to the end consumer through communication and sales channels. "The service marketing effort highlights the fact that the products to be promoted are immaterial and intangible, which makes them more difficult to execute" (SIMONI, 1997 apud MOTA, 2001, p. 134). In this way, it is understood that the characteristics of services make marketing actions more complex, since they cannot be seen, touched or stored. "Service companies would reach their end consumers more effectively if they exploited their differential against competitors, emphasizing the factors of brand prestige, reliability, promptness and quality of service (MOTA, 2001, p. 134)".

In this sense, Kotler (apud BOTELHO; COUTINHO, 2007, p. 3) defines that: "Tourism marketing is the set of activities that facilitate the realization of exchanges between the various agents that act, directly or indirectly, in the tourism product market". In this sense, tourism marketing can be understood as the application of marketing techniques to tourism products.

As Botelho and Coutinho (2007, p. 8) point out, tourism marketing "involves identifying market segments, promoting and developing the tourism product and providing tourists with information about the products on offer". From this point of view, the purpose of tourism marketing is to identify desires and needs, adapting the service and providing information about the tourism product.

In order to promote tourist destinations, marketing campaigns are used to publicize their attractions and services. Applying marketing techniques to tourist services is difficult because these products are heterogeneous because they are not uniform and intangible because they cannot be seen or touched. Mendonga Junior (2004, p. 32) states that: "tourist services are, for exactly the same reasons pointed out above for services as a whole, immaterial, intangible, non-storable and non-transportable goods".

> Given that the tourist product's intangible characteristics make it a unique element, made possible by the tourist trip, it can be seen that its commercialization requires specific marketing techniques, such as tourism marketing (MOTA, 2001, p. 143).

Thus, tourism marketing uses the tools of advertising and of advertising to promote services and products.

Historically, through research into advertising and the advertising, it can be seen that the great evolutionary milestone occurred in the means of communication that are responsible for propagation, and not in the foundations that underlie it. conceptualize (PERES, 2006). Thus, advertisements and publicity have had the media of communication as a way of spreading the word to a wider audience.

Propaganda and advertising are inseparable elements. The word propaganda derives from the Latin propagare, which designates reproduction by means of a vegetative multiplication technique (PERES, 2006, p. 40). This means that propaganda is an instrument for spreading various discourses. The term publicity comes from the Latin publicus, which means the quality of what is public; in other words, it means making a fact or idea public (PERES, 2006, p.40). In this sense, advertising is understood as a means of publicizing what is proposed, bringing it to the attention of the population. For Strocchi (2007), advertising is a mass communication because it is aimed at a heterogeneous population and its messages are transmitted through media tools.

For Strocchi (2007), one of the characteristics of advertising is the use of many symbols, because they are immediate and make it possible to communicate quickly and precisely. In this sense, we understand that advertising uses a wide variety of resources to effectively reach its target audience.

According to Fidalgo and Grandim (2005), symbols are signs in which, although there is no link of similarity, there is a conventional relationship between the representative and the represented. Insignias, emblems and also stigmata are considered symbols. In this way, it is understood that symbols can be found and used in the most diverse ways in everyday life. Pinto (2002, p. 17) also explains that,

> the enunciation device also involves the image, in texts where this semiotic system is present. There are few cases in which the only semiotic system present in a text is the image; the most common in contemporary media culture are mixed texts, which bring together verbal text and images, or verbal text and sound systems, or all three. Discourse analysis defends the idea that any image, even in isolation from any other semiotic system, should always be considered a discourse, rejecting the category of "iconic signs" or "icons" in which they are generally classified by semiologists.

Thus, Strocchi (2007, p. 144) states that "the consumer doesn't buy the product, but the image of it, with its symbolic images and representations". In this vein, it is understood that the product is sold by the image it conveys and by the symbolic aspects embedded in that product.

Propaganda and advertising always seek to use elements of everyday life in society to develop advertising discourses. To this end, Peres (2006) also reports that these are developed on specific dates in the annual promotional calendar, i.e. commercial dates such as Christmas, Mother's Day, Children's Day, among others, which can be called an Occasion or when unusual events occur, creating possibilities to promote campaigns, the Opportunity. It is therefore understood that advertising and publicity seek strategies to promote their advertisements, using the commercial dates included in the annual calendar and any events in society.

According to Peres (2006), advertising seeks to induce consumers to buy by showing that the product and/or service is capable of satisfying consumer needs. An essential component in fulfilling the purpose of advertising discourse is rhetoric, which would be elements inserted in these discourses that generate persuasion, leading to the consumption of the product or service. In this sense, advertising discourse aims to persuade by making it clear that the product or service can completely satisfy the consumer's needs, desires and dreams. In this way, Peres (2006, p. 55) also states that:

> With rhetoric it is possible to see theoretically which elements generate persuasion. The world of rhetoric is that of opinions, acting on human issues of a universal nature, and applies to any area of knowledge regardless of the technical particularities of science.

Thus, rhetoric is a technique used in advertisements to attract consumers. Through rhetoric it is possible to identify which elements are capable of persuading the target audience, as they deal directly with human relationships, leading them to make a purchase.

Advertising communication "is not interested in changing opinions, but in inducing purchasing behavior: its purpose, therefore, is to make people aware of and disseminate products or services and convince them to buy a certain product" (STROCCHI, 2007, p. 133). In this way, advertising discourse aims to persuade consumers to buy certain services and/or products.

All advertising discourse is made up of elements that make up messages, divided into: exordium, narration and epilogue. Peres (2006, p. 67) explains that: "In advertising discourse, it can be said that the exordium corresponds to the title of an advertisement, the first moment in which the public identifies with the advertising message". In this sense, it is understood that advertising messages have a beginning, middle and end, which together help to form the meaning of the advertisement they are intended to disseminate.

The narration is understood by Peres (2006) as the part that aims to describe the characteristics of the product or service offered, describing its qualities and differentials for the target audience. The epilogue is what closes the message: the enunciator must succinctly recap everything that has been said during the speech.

Tourism seeks to promote tourism products by creating images and aspects that are absorbed by both indigenous communities and potential tourists. Thus, the value of the tourist product will be defined based on the quality of the services offered during the visitor's experience at the destination. By knowing the value demanded by the customer, it is possible to achieve the highest quality of products and services, which leads to greater customer satisfaction with the product and services offered. In this sense, Mendonga Junior (2004) states that in the marketing of tourist services, two main factors compete to induce the consumer to buy the product: the attractiveness of the destination and the degree of trust that the customer places in the production of the tourist package.

CHAPTER 4

ANALYSIS OF PORTO SEGURO-BA ADVERTISEMENTS FROM THE PERSPECTIVE OF FRENCH DISCOURSE ANALYSIS (AD)

4. 1 FRENCH MATRIX DISCOURSE ANALYSIS (AD) AS A TOOL FOR UNDERSTANDING THE MEANINGS USED IN DISCOURSES

The historical context of French Discourse Analysis emerged: "In the theoretical context of France in the years 1968-70, at a time when the feeling of the limits and relative exhaustion of structuralism emerged" (BRANDAO, 2003 apud VIANA, 2011, p. 21). Thus, DA emerged as a means of analyzing the discourses of subjects.

French Discourse Analysis (DA) seeks to analyze the discourses used in the most diverse means of communication, in such a way that it evaluates how the discourse is used and why it was used. Discourse analysis aims to critically evaluate, explain and describe the communication resources used in the most diverse means of propagating language. Based on these considerations, Orlandi (2010, p. 15) states that:

> Discourse Analysis, as its name suggests, doesn't deal with language, it doesn't deal with grammar, although it is interested in all these things. It deals with discourse. And the word discourse, etymologically, has in it the idea of a course, of a journey, of running through, of movement. Discourse is thus a word in movement, a practice of language: with the study of discourse we observe man speaking.

In this way, it is understood that DA approaches discourse by covering all the factors inserted in the discursive context, critically evaluating the elements used in the discourse. Thus, Orlandi (apud VIANA, 2011, p. 25) reports that the conceptual model of DA,

> investigates how the symbolic is related to its historical-material context, in which it generates meanings or effects of meaning. To do this, it takes into account the ideological and symbolic elements that permeate language. AD is therefore concerned with language in movement (opaque and non-transparent) which involves subject and situation in a relationship between the unconscious and the social context.

Thus, it is understood that Discourse Analysis seeks to analyze the practice of language in the midst of the discourses in which the subjects are involved, taking into account the social context, as Orlandi (2010, p. 15) points out: "In Discourse Analysis, we seek to understand language making sense, as symbolic work, part of the general social work, constitutive of man and his history".

Discourse is inserted into people's daily lives and is present in the most diverse aspects of society, since every word is part of a discourse, as Viana (2011, p. 25) reports, "Discourse is totally involved in the networks that permeate the social, political, religious, economic and cultural labyrinths, with a popular and historically crystallized theme". As such, AD seeks to analyze the functioning of discourse, how this discourse was used, in what context and the subjects involved. Orlandi (2010, p. 10) points out that: "in short, Discourse Analysis aims to understand how a symbolic object produces meanings, how it is invested with significance for and by subjects". In this way, it is understood that Discourse Analysis sets out to understand language and its ways of signifying, the meanings produced in discursive processes, taking into account values, symbols, the subjects involved, aspects of society, among other factors that can somehow have an effect on the context of discourse.

In light of this, Orlandi (2007a, p.27) states: "Discourse is the effect of meanings between interlocutors". Discourse can be understood as a construction that takes place between two or more subjects in a given social context.

> [...] The construction of meanings between interlocutors takes place in the dynamic between paraphrase and polysemy. Paraphrasing is the movement of repetition that returns to the same thing, reiterating meanings. Polysemy is the movement of opening up to something different, to a new meaning. A discourse can tend towards paraphrase or polysemy, but it is always constructed under this tension between what is repeated and what is reinvented (BENETTI, 2009 apud VIANA, 2011, p.25).

Thus, discursive processes are established through the already said and the discourses that are still being formulated, in other words, between the old, the memory and the new. Orlandi (2010) points out that: there is an intrinsic relationship between what has already been said (past) and what is being said (present) this comparison can also be made in relation to interdiscourse and intradiscourse. It is thus understood that every word is related to other words, whether these words were said in the past or in the present.

One of the characteristics of discourse is the ambiguity of its use, since at the same time as it reveals ideology, it also hides it. Thus, the materiality of ideology is discourse and that of discourse is language (ORLANDI, 2007b). In this way, discourse has a multiplicity of meanings in its totality and has to relate them so that they can be understood, as (VIANA, 2011, p. 26) points out, "Discourse has the role of relating the multiple effects of meaning, but with a regularity that can be grasped. [...] It is linked to the representation that is constructed in social practices and their functioning." As such, it is understood that discourse includes a plurality of meanings that can be interconnected and formulated.

The analysis seeks to analyze the effects of meanings within the context of the discourses. In this way, we know that the basis of discourse analysis is the corpus. Orlandi (2010, p. 63) states that: "the construction of the corpus and the analysis are intimately linked: deciding what is part of the corpus is already deciding about discursive properties". Based on these considerations, we know that the corpus follows theoretical circles and that, when analyzed, it is fragmented into parts that allow it to be examined in order to understand the meanings it contains.

4.2. ANALYSIS OF PORTO SEGURO-BA ADVERTISEMENTS

The municipality of Porto Seguro is located in the extreme south of the state of Bahia, bordered to the north by the municipalities of Santa Cruz Cabralia and Eunapolis; to the south by Itamaraju and Prado; to the east by the Atlantic Ocean and to the west by the municipality of Itabela (MESQUITA FILHO, 2006).

Apart from the headquarters, Porto Seguro has the following towns and districts: Trancoso, Arraial d'Ajuda, Caraiva, Pindorama, Vale Verde and Vera Cruz, totaling a territorial area of 2,408.327, according to IBGE (2012).

Currently, the city's main economic activities are: tourism, which sustains the city's economy, since Porto Seguro's economy is anchored in this activity, plant extraction, agriculture, livestock and fishing (MESQUITA FILHO, 2006).

Investments in local tourism began with the cocoa crisis, when the government of Bahia set out to establish tourism as a source of income and jobs for the state. As Porto Seguro has many cultural and natural treasures, it was able to benefit from the state's economic policy (MESQUITA FILHO, 2006). In this way, Porto Seguro was able to develop economically as a result of the government's investments in tourism development policies, which promoted the many historical and natural riches found in the municipality.

As a result of the government's interest in tourism, the Bahia state government, through BAHIATURSA, zoned Bahia into seven areas for tourism exploration, one of which was the Costa do Descobrimento (MESQUITA FILHO, 2006, p. 47). This fragmentation of the regions allowed the government to plan its actions according to regional characteristics, making it possible to develop these places. Porto Seguro is therefore part of the Costa do Descobrimento along with four other cities: Eunapolis, Itabela, Santa Cruz Cabralia and Belmonte.

The city is considered to be Bahia's second destination and one of the most important in Brazil. It is rich in diversity and has many cultural and natural attractions. Throughout Brazil, 65 destinations have been classified by the Ministry of Tourism. In Bahia, five stand out and Porto Seguro is one of them, along with Salvador, Lengois, Mata de Sao Joao and Marau (SETUR, 2013). It is therefore known that in order to attract an increasing demand from tourists of different motivations, the destination has been seeking to format tourist itineraries highlighting not only the main attractions but also the local peripheral attractions.

The predominance of the sun and beach tourism segment, to the detriment of cultural tourism and other peripheral segments found in the region, coupled with a lack of planning, has minimized the development of the destination's potential, as well as the crucial attributes in the region's economic and social development, such as the lack of investment in education and the professional qualification of the population (MESQUITA FILHO, 2006). Thus, the municipality lacks strategic planning in its actions, both in terms of the essential attributes for the development of a region and in terms of issues related to tourism, which is one of its main economic activities, making it impossible to develop the region.

Among the cultural attractions found in the municipality are, in the upper town: the Historical Center, listed in 1973 by the National Historical Patrimony, which is the former Pago Municipal (or town hall and jail, 1756), where the town museum is housed, the Church of Nossa Senhora da Pena (1535), the Marco da Posse (1506), the Church of São Benedito (1549), and the Jesuita College. In the Lower Town you can also find: the First Church, already in ruins, the Memorial to the Discovery of the Epic, the Jaqueira Indian Reserve and other attractions. Natural attractions include the Recife de Fora National Park, the beaches and the Atlantic Rainforest. The city has a large number of tourist facilities such as hotels, inns, resorts, restaurants and bars. In order to promote the destination's tourist attractions, the region's public and private authorities use marketing resources such as advertisements to increase tourist demand in the region.

The aim was to analyze the discourses inserted in cultural tourism advertisements in the municipality of Porto Seguro (BA), given its importance for tourism and the fact that it is one of Bahia's five tourism destinations. In this way, we analyzed media such as the internet, as well as folders, videos and photos that promote the destination's cultural attractions.

The advertisements analyzed include in their discursive corpus elements of local culture in which they promote cultural tourism in the region, with the aim of strengthening this segment and attracting more and more demand.

The Porto Seguro Institutional video released by the city's Tourism Department in 2011, which lasts 4:09, can be found on media such as the internet on the website of the city's

Tourism Department, and shows the format of this tourist destination, with the aim of spreading the word and attracting more and more people to visit the city. In this sense, according to Peres (2006), this advertisement was designed to promote the attractions of the municipality of Porto Seguro, showing that it will be able to satisfy the desires and needs of the target public.

Initially, the video features the image of the Indian, a symbol of the discovery of Brazil. The indigenous language is explored in order to convey the impression that they are really in a jungle, far from living with other ethnicities and cultures and the transformations to which they have been exposed. This image is accompanied by the phrase "four destinations in one place", the exordium of the advertisement, which tells us that Porto Seguro is a destination where you can find variety, where routine and idleness are far away, where there is much to see and explore. The fauna and flora are also elements that are intensely focused on, with the aim of attracting people to the new, exotic and natural. In this way, the video features 4 different tourist segments, since Porto Seguro-BA is a diversified destination. The beginning of the video is characterized as the title of the advertisement, which is the introductory message according to Peres (2006).

Throughout the video, the "object of persuasion" revolves around the image of the Indian, who is shown as if he still retained his original identity, his customs still preserved, his musicality, his clothing, his language, his paintings, his musical instruments and his shit, with the aim of attracting tourists to see the "wild", the "out of the ordinary", what they are not used to seeing on a daily basis. That's why tourists want to experience the new, to get away from everyday life. In this conjecture, rhetoric is identified as an element of persuasion, inducing the public through an image that has already been deconstructed, as Strocchi (2007) points out when he says that advertising communication must build and format an image that is acceptable and different from the others, making that service or product unique. In this way, the advertisement must propose an image that stands out from the rest and is accepted, attracting and satisfying the needs of the public.

Figure 01: Institutional video of Porto Seguro, BA, 2013.

Source: www.youtube.com.br

Next, it slightly presents the main points of the Historic Center as the place that tells the story of Brazil, where it all began. This attraction does not have an in-depth focus, like the image of the Indian. The narrative of this advertising message, according to Peres (2006), seeks to describe the attributes of the municipality by promoting its history, its qualities and the fact that it was the birthplace of Brazil.

The beaches, one of the most popular attractions for tourists to Porto Seguro, are also featured, showing the coral reefs and leisure activities. Caraiva, Trancoso and Arraial d'Ajuda are also shown, highlighting sun and beach tourism and religious tourism. This section shows one of the most visited segments, characterizing mass tourism in the municipality. This type of tourism doesn't have as much focus either, and is shown briefly in the video.

Gastronomic tourism is also one of the city's attractions, with its changing colors and tropical fruits. At night, as a place to visit, it shows the "passarela do alcool" and its options, which are usually very popular during this time of year, especially in high season. This interval in the video is characterized as the end or epilogue, as Peres (2006) points out, briefly returning to the entire discourse throughout the advertising video.

Throughout the video, the soundtrack features the musicality of the Indians and all the formatted scripts are shown within an indigenous context. The video displays a wealth of colors, such as red, which symbolizes passion and joy; green, which symbolizes the forests and tranquillity; yellow which represents the riches of the place, orange which expresses energy and joy, blue which refers to the beaches and plenitude, always in warm tones, this predominance of vibrant colors and the varieties of them become very inviting motivating the tourist to visit Porto Seguro, injects in the tourist the euphoria, the curiosity to know a place so diverse and cheerful, where it can vary from the rustic to the refined, in addition to passing the image of Porto Seguro being a true natural and untouchable paradise. According to Orlandi (2009), discourse does not seek the true meaning, the truth inserted in the context, but the real meaning in what governs linguistic and historical materiality. Thus, ideology is not apprehended, the unconscious is not controlled by knowledge. Language itself functions ideologically, having its materiality in this game.

Next, we analyzed the advertisement "Eu Nasci em Porto Seguro. My name is Brazil", which is located at the entrance to the city on the BR-367 access road. The speech: "I was born in Porto Seguro. My Name is Brazil", makes an analogy with the Discovery of Brazil and emphasizes the importance of Porto Seguro as the birthplace of the country, which gives tourists the sensation of returning to the past and stimulates the visitor's imagination, as well as arousing curiosity about getting to know the place where the history of Brazil began up close and personal, providing a feeling of welcome, this sensation arises from the way the narrative is presented, the ideological position. The use of white, chosen to emphasize the statement in black, has a very attractive effect and makes a strong impact. Fidaldo and Gradim (2005) point out that symbols are associated with colors. Each color has its own representativeness. The image of the Indian with variations of hot and cold colors is inserted in the advertisement, giving even more emphasis to the statement, as well as conveying the image of an exotic place. Orlandi (2009), as already mentioned, says that meaning is determined by ideological positions in the socio-historical process of society and does not exist in itself, so the ideological positions of the discourse give the reader an idea of proximity to the destination.

It is understood that discursive memory is capable of changing the meanings embedded in discourse, as it supports the discourse being formulated.

Figure 02: I was born in Porto Seguro. My name is Brazil, 2013.

Source: Own elaboration 2013.

We then analyzed the Porto Seguro-BA Copa 2014 advertisement, which was created by the municipality's public and private authorities and can be found on the website of the Porto Seguro-BA Tourism Department. It seeks to take advantage of the 2014 World Cup in Brazil and features a transience of colors in an attempt to attract the public's attention. The white background used served as a subsidy so that all the other colors used could be reflected and highlighted. In this context, the occasion of the World Cup was used to publicize the attractions of Porto Seguro-BA. Peres (2006) says that advertising and publicity can use occasions such as commercial dates to develop advertisements.

Figure 03: World Cup, 2013.

Source: http://blogdomiolobaiano.blogspot.com.br/2012/04/e ntidades-de-porto-seguro-se-alinhar.html

On the left, the Porto Seguro logo created for the advertisement features a multiplicity of colors: green, which symbolizes the natural and cultural characteristics that are intensely found in the region, and orange, which brings out the city's joy and energy. Sun and beach tourism is represented in blue and orange in the form of symbols around the words: "Porto Seguro is much more!", which tells us that Porto Seguro is a place where you can find variety, an infinity of motivations. As Pinto (2002) points out when he says that the enunciation passes through the image, that is, the image transmits messages, where meanings are inserted. For him, there are few advertisements that have only the imagery, and in this sense advertising has mixed elements, bringing together the verbal and the non-verbal. Discourse analysis, in this context, identifies the image as a discourse full of meanings. Strocchi (2007) explains that the consumer is attracted by the image represented by the product and the symbolic images and representations inserted.

In the central part of the advertisement, the logo of the World Cup, the main theme, presents a mixture of warm and cool colors such as orange, red, yellow and green which refer to the energy, intelligence, strength, riches of Brazil, the passion for soccer experienced by the Brazilian people and the biodiversity considered to be one of the richest in the world with a multitude of natural landscapes. As Fidalgo and Gradim (2005) point out, colors can be classified as warm or cold. As a discourse narrative, the advertisement features the statement: "FIFA WORLD CUP", which means World Cup. As an epilogue: "Brazil", it complements by announcing the place where the World Cup will take place. The advertisement's essence is the Brazilian people's passion for soccer, in an attempt to attract more and more tourists to Porto Seguro, since the World Cup will take place in Brazil.

Based on these considerations, it can be understood that advertising communication plays the role of formulating the product's image, with a view to its acceptance by the public.

The Porto Seguro logo, on the other hand, is a symbol found in a variety of advertisements, announcements and so on, and was developed by the municipality's City Hall together with other public bodies. In its presentation, the image of the Indian emerges as a strong representation of the city, since the Indian is the symbol of the Discovery of Brazil. In this sense, Strocchi (2007) says that advertising makes use of symbols because they are direct and contiguous, promoting fast and effective communication.

Figure 04: Porto Seguro logo, 2013.

Source: http://tvmaanaim.hd1.com.br/

The logo features a variety of warm and cool colors such as black, yellow, red and variations of green that make an impact, attracting the public's attention and highlighting the city's cultural and natural characteristics. The leaves with variations of green indicate the plurality of natural landscapes found in the city, the red and yellow the strength and intensity of the indigenous culture, and the black emphasizes the indigenous traits strongly present in the population. In this context, the logo of Porto Seguro-BA brings into the discursive corpus elements that are already consolidated, the already said, keeping in touch with the past, as Orlandi (2010) discusses.

The predominant color of the "Porto Seguro" discourse is red, which creates the sensation of energy, warmth and happiness. The white background serves to highlight and reflect the other colors used in the advertisement. Fidalgo and Gradim (2005) report on the effect of color on meaning, which provokes sensations and feelings.

The epilogue: "This is where Brazil was born" makes an analogy with the discovery of Brazil in an attempt to emphasize that the city is the birthplace of the country. The end of this advertisement takes up the historical and cultural elements imprinted in the meaning effect.

According to Orlandi (2010, p. 43), "words speak to other words. Every word is always part of a discourse. And every discourse is delineated in relation to others: present sayings and sayings that are lodged in memory". As such, it is understood that words are elements that communicate, weaving discourses and effects of meanings that may be inserted in past processes and which are also constructed with new sayings.

CHAPTER 5

FINAL CONSIDERATIONS

After critically examining the advertisements, analyzing the discourse-effects of production and reaction used in the discourses of tourism in Porto Seguro-BA through Discourse Analysis, we can see how the advertising discourses work in the municipality's advertisements, understanding the effects of meanings inserted.

Thus, based on the analysis carried out using French Discourse Analysis (DA) on the advertisements surveyed in the area in question, the corpus of the discourses employed can be evaluated, in which the constant use of the image of the Indian as an "object of persuasion" can be seen, with the aim of awakening in the visitor an impression of the new, exotic and natural, and of being in a jungle far from living with other ethnicities and cultures and the transformations to which they have been exposed since the advent of globalization and technological advances in society.

The use of the image of the Indian, highlighting his cultural identity, his musicality, his clothing, as if he still retained his authentic identity, gives the impression that Porto Seguro is a wild, exotic, liberated place, different from what visitors are used to seeing, a place where anything can happen. In this way, the use of this symbol of local cultural identity in advertisements is intended to arouse curiosity and imagination, with the aim of attracting more and more visitors.

In the context of the municipality's advertising, analogies are constantly made to the Discovery of Brazil, emphasizing the city's connection with the country's early history, strengthening the didactic-historical-ideological conception that Porto Seguro is the birthplace of Brazil.

During the analysis of the Porto Seguro advertisements, it was possible to see that the most diverse dimensions of tourist activity such as sun and beach, ecotourism, gastronomy, festivals, and various other potentialities such as cultural tourism are integrated into their discursive elements, arousing the interest of the most varied tourist profiles, thus incorporating a greater number of tourist segments into the destination, such as those shown in the Porto Seguro Institutional video and observed in the advertisements analyzed.

In addition, advertisements use a variety of colors, usually warm colors that emphasize joy, diversity and are linked to different meanings in an attempt to attract attention and more visitors.

In this way, the speeches in the Porto Seguro advertisements give visitors the feeling of being in a true paradise, being able to reaffirm their cultural identity, live intensely, have contact with the different, new, exotic, escape from their daily lives, live without rules.

We can therefore conclude that Porto Seguro's advertisements are instruments for promoting the municipality, publicizing its cultural and natural attractions. In their discursive corpus, they use identity elements, the transience of colors and a multiplicity of meaning effects, with the aim of increasing tourist demand.

We hope to contribute to the scientific community with relevant scientific data for future research projects in the area, participation in cultural and scientific events, generating basic knowledge for future research and intervention projects in the area of tourism and culture. Science as a form of accumulative and dialectical knowledge is growing and becoming an important human dimension that helps in socio-cultural realization.

6. REFERENCES

ALBERTO, Diana Priscila Sa; OLIVEIRA, Karla Cristina Damascene de. **"E FOLIA DA ILHA, E FOLIA DO SANTO"1: TURISMO CULTURAL E A FESTIVIDADE DO GLORIOSO SAO SEBASTIAO DE CACHOEIRA DO ARARI - ILHA DO MARAJO/ PARA**. CULTUR - Culture and Tourism Magazine - year 03 - n. 02 - April/2009 - Special Issue 2009.

AVILA, Marco Aurelio. Politics and planning in culture and tourism. Ilhdus: Editus, 2009.

BARRETO, **Margarita. Tourism and Cultural Legacy: The Possibilities of Planning.** Campinas, SP. Papirus, 2000 (Tourism Collection).

BENI, Mario Carlos. **Structural analysis of tourism. 8ª ed. Atual. Sao Paulo**: Editora Senac Sao Paulo, 2003.

BESSA, Dante Diniz. **Theories of communication. Brasilia**: University of Brasilia, 2009.

BOTELHO, Loreta Cabral; Coutinho, Helen Rita Menezes. **Tourism Marketing in the City of Manaus**. In: Revista Eletronica Abore. Amazonas. Issue 03/2007. Available at: < http://www.daneprairie.com>. Accessed on: 15 Fer, 2013.

BRAZIL, Decree-Law No. 4. 340, of August 22, 2002. **Establishes principles and guidelines for the implementation of the National Biodiversity Policy**. Available at:< http://www.mma.gov.br/port/conama/legiabre.cfm?codlegi=363>. Accessed on: April 15, 2013.

WORLD CUP. Available at:<http://blogdomiolobaiano.blogspot.com.br/2012/04/entidades-de-porto-seguro-se-alinham.html>. Accessed on: May 20. 2012.

COSTA, Diogo Menezes. **Heritage archaeology: thinking about building**. Habitus Magazine, Goiania, v.2, p. 333-360, 2004.

COSTA, Flavia Roberta. **Tourism and Cultural Heritage: interpretation and qualification.** Sao Paulo: Editora Senac Sao Paulo: Edigoes SESC SP, 2009.

DIAS, Reinaldo. **Planejamento do turismo**: politica e desenvolvimento do turismo no Brasil.Sao Paulo: Atlas, 2003.

DIAS, Reinaldo. **Sociology of Tourism**. Sao Paulo: Atlas, 2003.

DIAS, Reinaldo. **Tourism and Cultural Heritage - resources that accompany the growth of cities**: Saraiva, 2006.

FRANTZ, Michelle Bencciveni Franzoni et. al. **Marketing mix analysis in a franchise system.** VII SEGeT - Simposio de Excelencia em Gestao e Tecnologia. 2010.

FIDALGO, Antonio; GRADIM, Anabela. **Manual of Semiotics.** UBI. Portugal. 2004/2005.

GANDARA, Jose Manoel Goncalves et. al. **SEEDS FROM THE ATLANTIC FOREST: CONFORMATION OF THE CULTURAL PRODUCT FOR THE DESTINATION ITACARE - BAHIA.** CULTUR / ANO 5 - N° 01/SPECIAL - JAN (2011).

GEERTZ, Clifford. **The interpretation of cultures.** Rio de Janeiro: LTC, 1989.

GRANDISOLI, Joao Luiz. **Marketing Mix or Marketing Compound.** Available at:<http://www.portaldoMarketing.com.br/Artigos/Marketing%20Mix.htm>. Accessed on: May 25. 2013.

IPHAN. **National Historical and Artistic Heritage Institute.** Available at:< www. Iphan.gov.br>. Accessed on: June 20, 2013.

KOTLER, Philip. & KELLER, K. L. **Marketing Management: The Marketing Bible.** Sao Paulo: Prentice Hall. 2006.

KOTLER, Philip. **Marketing Management.** Bazan Tecnologia e Linguistica translation; Arao Sapiro technical revision. 10ª Edition, 7ª reprint. Sao Paulo: Prentice Hall, 2000.

PORTO SEGURO LOGO. Available at:< http://tvmaanaim.hd1.com.br>. Accessed on: May 10. 2012.

LOHMANN, Guilherme. **Tourism theory: concepts, models and systems.** Sao Paulo: Aleph, 2008.

MARCONDES FILHO, Ciro. **Para entender a comunicação: contatos antecipados com a nova teoria.** Sao Paulo: Paulus, 2008.

MARQUES DE MELO, Josd. **Social Communication: theory and research**. 4. ed. Petropolis: Vozes, 1975. 300 p.

MENDONQA JUNIOR, Erico Pinto. **Marketing and Competitiveness in Tourism in Bahia**. Salvador: Department of Culture and Tourism, 2004.

MESQUITA FILHO, Odilon Pinto. **Tourism in Porto Seguro: aspects**. Itabuna; Ilheus: Via Literarum, 2006.

MINISTRY OF TOURISM. **Cultural tourism:** basic guidelines. Brasilia, 2008.

MOTA, Keila Cristina Nicolau. **Tourism Marketing: Promoting a Seasonal Activity**. Sao Paulo: Atlas. 2001.

UNWTO. **Introduction to tourism.** Translated by Dolores Martin Rodrigues Corner. Sao Paulo: Roca, 2001.

ORLANDI, Eni Puccinelli. **Discourse analysis: principles and procedures**. 9ª Edigao, Campinas, SP Pontes Editores, 2010.

. **Discourse Analysis**: principles & procedure. Campinas: Pontes, 2007a. . **As formas do silencio**: no movimento dos sentidos. Campinas: Editora
UNICAMP, 2007b.

ORLANDI, Eni Pulcinelli. **Language and how it works:** the forms of discourse. Campinas: Pontes, 1996.

PERES, SERGIO. **Advertising at COPA: An analysis of the advertising discourse at the World Cup.** 1st Edition, Campinas, Editora Komedi, 2006.

PORTUGUEZ, Andreson Pereira. **Tourism, memory and cultural heritage**. Sao Paulo: Roca, 2004.

SPINOLA, Carolina de Andrade. **Post-Modern Tourism - A Paradoxical Context**. 2008.

STROCCHI, Maria Cristina. **Psychology of Communication: Manual for the study of advertising language and sales techniques.** 1st Edition, Sao Paulo: Paulus, 2007.

UNESCO. **United Nations Educational Organization**. Available at:< www.UNESCO.gov.br>. Accessed on: June 20, 2013.

VALLS, J. F. **Destino turistico. In: Gestion de Destinos Turisticos Sostenibles.** Barcelona: Gestion 2000, 2004.p. 17-45.

VASCONCELOS, Luciene Ricciotti. **Integrated communication planning: a survival manual for 21st century organizations.** Sao Paulo: Summus, 2009.

VIANA, Moisbs dos Santos. **LABIRINTS OF IDENTITY: Analysis of the Identity Discourse of the Ternos de Reis Groups in Itapetinga-Bahia.** Master's dissertation in culture and tourism. UESC. Ilhbus, Bahia, 2011.

INSTITUTIONAL VIDEO OF SAFEPORTS-BA . Available at:<www.youtube.com.br>. Accessed on: December 15, 2013.

Printed by Books on Demand GmbH, Norderstedt / Germany